O M

1. Yogi Vithaldas

THE YOGA
SYSTEM OF HEALTH
AND
RELIEF FROM TENSION

by

YOGI VITHALDAS

BELL PUBLISHING COMPANY, INC.

NEW YORK

This edition published by Bell Publishing Co., Inc.
a division of Crown Publishers, Inc.

s t u v w x y z

CONTENTS

ILLUSTRATIONS

Illustrations

*All the photographs in this book are reproduced by
courtesy of Beiney, Photographer, London*

YOGA AND THIS AGE

Chapter I

YOGA AND THIS AGE

The Yoga system of health is a culture that has been practised by the yogis in India for thousands of years. Its roots lie buried deep in the past, but its message is addressed no less surely to the people of to-day, living in the restless atmosphere of the modern world. Yoga lays stress on bodily and mental poise and produces an equanimity of spirit that is most beneficial to the whole nervous system. It trains the student in the basic principles of health, and creates a true placidity of nature that allows great intensity of activity of both mind and body, when such activity is necessary.

The question is whether the Western world needs Yoga and is ready to attach to it the significance the East attaches. The Westerner must perforce admit that the modern world is one of agitation and nervous tension. Does he get along satisfactorily or does he just 'muddle through'? Life expectancy has certainly gone up in the last hundred years, but this has not been due to an innate development of resistance to disease, but rather to the increase in medical knowledge. Sir Farquhar Buzzard, Physician in Ordinary to the King, maintains that in England one person in every fourteen suffers from nerves to such an extent as to warrant treatment, whether it is given or not and that one million weeks of working men's time are lost every year because of nervous disorders alone. We have further his authoritative statement that one-third of all the ills that man is heir to in these days are due to nervous disorders, and are not organic breakdowns.

The need for Yoga therapy, which deals so exhaustively with nerves and their effects, is thus seen to be very real, for ill-nourished and uncontrolled nerves sap the vitality of a nation, and affect its physical condition and its mental outlook. Yoga is not advocated to teach the Western city dweller the Indian rope-trick or any other abnormal practice (and among these false ideas must be included the use of Yoga to effect indefinite prolongation of life), but to expand his own latent powers, physical, mental, and spiritual, to their fullest possible extent.

The life of the average man, and by that I mean the majority, is a biological span that he inherits from his parents and grandparents, but many people, by paying too little attention to the fundamentals of living, literally fail to get the most out of life, and die before the completion of their inherited life span. Yoga, as taught in this book, will show how the longest life may be attained.

Few of us so conduct our lives as to exhaust our span of years, and most of us are ignorant of what treatment of the body is conducive or detrimental to the prolongation of the inherited tissues.

To-day only one person in a hundred thousand lives to celebrate his or her hundredth birthday, and of those the greater proportion are women, who, because of their sex, are spared the more arduous occupations that fall to the lot of the male. The reader has an even chance of seeing just over sixty years, and his chance of survival is double what it was a hundred years ago.

Yoga builds stamina, not strength, and calls for moderation in eating, drinking of alcohol, and especially in smoking, and above all demands a happily balanced frame of mind. Statistics show that long-lived people are those of contented mind, for the mind does not grow old with the body if the right precautions are taken and if the blood-supply to the brain is maintained. This is common knowledge to every doctor, and the Yoga system allows for this and has exercises for keeping the brain supplied with blood.

There is nothing strange or unnatural in this. In the dim past, when the progenitor of man trotted about using his hands as supports, much as apes do, his head was lower than his heart, and gravity sent the blood rushing to the brain with comfortable ease, and kept it healthy and responsive during the process of evolution. Walking upright, as we do now, gravity does not send this blood to the brain, and the heart pumps up a reduced quantity through the carotid arteries. The outflow is through the **jugular** veins, which are larger and take off more than the carotid can get up with ease. The heart thus has demands made upon it that it cannot adequately fulfil, and Yoga teaches the obvious remedy, i.e. an exercise that puts the student standing on his head. This, however, is only one of many important exercises, and Yoga must not be considered as specializing in an adaptation of the individual to the gravitational pull.

Hatha Yoga, in fact, re-educates the abdominal muscles, buttocks, and lumbars to resist the gravitational pull, keeping the body in a correct posture and the intestines in the pelvic cavity. As to standing on the head, what at first sight would seem to be a childish exercise is really an obvious method of countering an inadequate blood-supply to the brain. Such an inadequate supply results in degeneration of the brain tissues, and in a loss to the mentality of the alertness necessary to cope with the trying conditions of modern civilization. Experiments to prove that the blood-flow to the brain affects mental output have been successfully carried out, and it is common knowledge that when the stomach is overfed the blood is drawn from the brain and rushes to the abdominal regions to cope with the overtaxed stomach. A light meal for the mental worker means less blood for the stomach and more for the brain, with a consequent improvement in mental efficiency. It is in fact probable that the Sunday dinner, deflecting blood from the brain to the stomach, leaves it 'fagged out', and produces the well-known 'Monday morning feeling'.

16

Yoga and this Age

Yoga has exercises stimulating the whole of the abdominal muscles, the spine, and the lungs. That attention should be given to these parts, instead of to building up firm muscles on biceps and legs, is not an Eastern fad, but is based on medical common sense. Doctors studying old age in Prague saw that in the aged there is the deterioration of just those parts of the body that the yogic exercises keep virile and supple. Yoga has taken elementary scientific knowledge and adapted it to a specialized system, and in this light it is realized that there is nothing mysterious in Yoga.

The brain and the spinal cord also have every attention from the yogi, because the yogi knows them as the two great centres of the whole nervous system of the body. The effect of the nerves on the body cannot have its importance exaggerated. Dr. G. E. Hall, of Toronto, explains that a stimulated nerve fibre oozes a secretion called acetylchaline, which, in excess, is poisonous and has been proved by experiment to create ulcerated stomach and some diseases of the heart. Anger, which is a strong emotion, sends its messages from the brain, down the nerves, and stimulates the nerve endings to discharge acetylchaline in excess. This poisons the body, and the result of anger is low spirits and misery. Is it, then, the absurd invention of a weird cult to insist on the composed body and mind? It is nothing more than the practical application of common sense.

The reader may consider himself composed, but might

17

this not be due to nervous re-actions being repressed in his conscious life? When he is asleep the conscious is off duty, the guard is gone. The average sleeper, without knowing it, of course, changes his position between twenty-five and fifty times a night. That is no composure of body and mind.

The rhythm of the breathing exercises is reflected to some extent in the accepted scientific fact that the brain generates electricity under the influence of rhythmic chemical action, whereby the thinking process is carried on, and other bodily functions. Dr. T. J. Case, of Chicago, says that the nerve-cells pulse electrically at about ten beats per second, and a break of rhythm means that there is some affliction at the area where the rhythm is broken. That rhythm is fundamental to life is indicated in many ways, as for instance in breathing and in the love of music and dancing. Dr. Rex B. Hersey has charted a man's emotions, and finds that they run in cycles of crests and depressions over an average period of four to five weeks. Yoga recognizes nature's demand for rhythmical living and trains the student accordingly.

That Yoga believes so strongly in the potentialities of the mind can readily be understood when it is realized that the subconscious, while giving evidence of phenomenal powers, is not yet by any means understood. Hypnotism, one method of exploring the subconscious, has revealed that the subconscious absorbs things we have never dreamt came within our ken, and when stimulated it will reveal them in the most minute and accurate details. Working through the conscious, however, its hidden knowledge is often revealed in distorted shape, to the individual's confusion, much as dreams juggle with the impressions left by the day.

The hidden mind profoundly influences the conscious mind, so Yoga strives to understand the subconscious.

The effect of the mind on the body is being established with increasing force by the progress of medical science, and many statements in this book are corroborated by authoritative statements of fact made in *The Force of the Mind*, by Dr. A. T. Schofield, M.D., M.R.C.S., an admirable work, to which I am deeply indebted.

As to the influence of the mind on the body, we have Sir James Paget writing to Sir Henry Acland in 1866 that one of his patients might surprise the whole medical profession by being relieved of her maladies were there some force of assertion which would create the will to bear, forget, or suppress the turbulences of the patient's nervous system.

Modern medicine can no longer afford to ignore those potentialities of the mind that Yoga has always recognized, and that it has harnessed in a course of real therapeutics. Yoga recognizes that man is psychical and physical, and intermingled as these two aspects are, endeavours to systematize and establish the psychic laws in relation to the physical. It is dawning on the East that to impress the West she must systematize her knowledge; for the Westerner, unlike the Eastern student, will not blindly follow in faith and rely on intuition that he is in the hands of the right teacher. It may be this lack of intuitive perception that has prevented the West assimilating much the East is ready to give, but which it cannot give if the West questions its authority. Now that the knowledge of the East is being formulated on Western lines, Western science feels more inclined to accept and understand, and even to adapt Eastern wisdom to its own needs. This continuity of mind and body is shown in ways which we tend to pass over, but Sir B. W. Richardson indicates for us the common association of the emotion of love with the heart, irritability with the liver, and, significantly enough, the highest point of mental insight with the stomach—the solar plexus of the sympathetic system. Significantly, the great cosmic energy of the yogi, called Kundalini, is described as resting in the abdominal regions, and, when stimulated, gives supermental powers.

Hack Tuke, in his classic on the subject, gives us figures in respect of the power of the mind on the body, stating that 36 per cent of the effects on the vascular system comes from the intellect; 56 per cent on the glands and organs of the body comes from the emotions; and 8 per cent on the so-called voluntary muscles comes from the will. Yoga is a science confident in its knowledge that no mechanical system could adjust itself to the constantly changing environment of the body unless there were a flexible centre such as the mind in control. This is confirmed by Professor Clouston's statement to the Royal Medical Society in 1896 that the mental aspect is a factor for good or evil in any disease.

Religion, calling for ethical living, links up with natural science, for if the mind and body are so linked that the mind can influence the body, it is equally probable that the body can affect the mind, and immoral use of the flesh or its abuse can warp the mind. Medical men have compiled frightening lists of diseases influenced or created by the mind, from cancer and heart diseases to simple stomach-ache. The mind can induce paralysis, and death may be caused either by a natural physical reaction in the body, or by hysteria where non-existent diseases or conditions are imagined.

Just think of the power of the mind! It holds agony and death in its grasp, and yet science is only just taking it seriously as a factor in therapeutics, whereas Yoga has been striving after an understanding for two thousand years. Does not this demand that the West should seriously consider a culture with such a tradition?

The mind also has a curative effect when directed and applied in the right channels, and we have only to think of hypnotism, faith, hope, prayer, confidence, will-power, auto-suggestion, and religious mysticism to realize the power of mind over matter. These aspects of the mind have in practice saved people from years of prostration, pain, and from death, and they cannot be ignored. Yoga sees this as an indication of the power of the mind, and

21

directs it accordingly by a system based on knowledge from prolonged study over centuries.

The yogi endeavours to create a life force within himself, and Western science itself acknowledges something much akin in the *vis medicatrix naturae*, a natural inherent combative force in function against disease, described by Dr. A. H. Carter as a 'working out of the balance of nature', that harmonious system of life referred to earlier in the chapter.

Captain J. A. Hadfield, M.A., M.B., in his contribution to that excellent work, *The Spirit—God and His Relation to Man*, states: 'Whatever their ultimate origin, there are resources of power whose existence we do not ordinarily recognize but which can be made available for the purposes of our daily life.' He instances genuine cases of feats done by people under unusual circumstances and strong emotions which could not be done afterwards in the ordinary course of living. Case after case is quoted, revealing in all that the subject has made access unconsciously, or rather subconsciously, to hidden sources of power, and afterwards suffered no fatigue from his extraordinary expenditure of energy. Under hypnosis cases before Captain Hadfield exhibited abnormal feats of strength. This leaves no alternative but to see the will as a power in itself governing the whole body, and the emotions as providing impetus to the will.

A governing law seems to be that progress comes from the adaptation of the individual, with all his special powers, to the prevailing social life, and it is this that forms the basis of the moral outlook, for as Captain Hadfield points out, power used to the detriment of the community is eventually reflected back on the individual who abuses it, with unhappy consequences to himself.

I shall end my chapter with another quotation from this interesting book: 'This art of resting the mind and the power of dismissing from it all care and worry is probably one of the secrets of energy in our great men.'

That is worth a thought in an age that saps mental energy and nervous strength when it is most needed.

YOGA THERAPY

Chapter II

YOGA THERAPY

Yoga is an ancient and extended system of psycho-therapy. The points of difference between the psycho-therapy of the East and that of the West are the emphasis laid by the East on body poise, breathing, and the objectification of the body.

The spiritual aspect of Yoga is only in part frowned upon by psycho-analysts, as there is a school of psychologists who accredit their cures to the establishment in their patients of a renewed sense of well-being and of confidence in God.

That Yoga is a system of psycho-therapy created over two thousand years ago and successfully practised ever since, with amazing results, is a fact that brooks no argument; that Yoga's system should be dissected and adapted for the Western world and practised by Western psychologists is my fervent hope.

The ground that can be covered by a psycho-analyst in respect of his patients includes all anti-social habits, such as acute timidity, temper, hysteria, depression, sex com-

27

plexes, etc., and physical disorders such as constipation, indigestion, and irregular functioning of the processes of excretion and menstruation, and mild skin diseases. That patients come eventually to the psycho-analyst is because medical science has failed to eradicate what are, in so many cases, fundamental ills. This is not to infer that the medical profession never achieve satisfactory results in such cases. On the contrary, the doctors cure with consummate skill the majority of sufferers, but they fail lamentably with some who apparently suffer from the same functional disorders as the majority they have cured. Medical science, with its present curative devices, must then step aside for the psycho-analyst, a man whose specialized knowledge is ever becoming more frequently sought by those who are in danger of succumbing to the unnatural circumstances of modern life.

All this, though it is well in the domain of psychotherapy, does not constitute the main urge for the existence of Yoga-therapy. The main urge is one that Western psycho-therapy fights shy of absorbing, yet it is more general than any functional disorder; it is the urge for that knowledge whereby the individual may be brought to a full realization of his true self and his vast potentialities.

Few people can have the sincere conviction that they are exploiting to the full their physical and mental attributes, and the knowledge of this works insidiously throughout all they endeavour to accomplish in life. Half the sterile lives in the world are due to the unconscious (or sometimes conscious) realization of the inadequacy of the individual to reconcile his life with what he feels should be the true method of living. The result is a sense of frustration, for the individual wants something which he cannot identify. A knowledge of what a man wants is gained only after searching self-analysis, and the true Self is only revealed when its potentialities are made apparent. The postures, breathing exercises, and objectification of the body induced by Yoga-therapy automatically bring into being a state of mind perfectly attuned to the world around. In other words, Yoga makes the patient his own psycho-analyst, and it is from the study of Yoga-therapy itself that a satisfactory system is cultivated, with its consequent poise and peace of mind.

The nervous diseases reflected in so many disorders of the body and mind are the outcome of straining and tension of the faculties in a conscious or unconscious effort of the individual to adapt himself to his environment. The psycho- and Yoga-therapeutists endeavour by their own particular methods to relieve this tension and substitute a true relaxation of the mind and body disciplined to work harmoniously. The psycho-therapeutist discriminates from the flowing thoughts and ideas released in the main from the subconscious and applies those that will affect beneficially the ailment of the patient. In the majority of cases the application of these thoughts is the means of bringing about relief from repression in the subconscious. The Yoga-therapeutist believes, not without reason, that, as his own psycho-analyst, he is better qualified to apply the most effective method for his own betterment. The psycho-therapeutist, on the other hand, condemns a system whereby the cure of a diseased mind has to be undertaken by that same mind. It is here that the ways divide, for the yogic student believes in a special quality of the mind whereby he can objectify it and rectify its prevailing weakness.

The objectification of the mind is made possible by the mind's functioning on varying levels of consciousness. Three of these levels are clearly differentiated—the physical (cognition of the world around us); the emotional (passion, anger, greed, etc.); and the mental (intellectual). Freud, the great psycho-analyst, has mapped out the mind into three levels—conscious, pre-conscious, and sub-conscious. The Indian mind accepts with ease, as an axiom, the extension of planes of consciousness; the Westerner in the majority of cases has difficulty in accepting so alien a conception. The yogi does not postulate a different mind for each level of consciousness, but one mind capable of working at different levels. To retain consciousness of the mind at one level and see it working on another allows, if necessary, mental observation of any distorted activity. It is obvious that one cannot get accurate powers of mental perception from the lowest level of a mind working on a higher plane, so the yogi raises his mind to its highest level in order to study its activities on lower levels of consciousness. Yoga aims at that command over the body and mind that is the state of individual existence in which the mind, working smoothly in a physically fit body on a high level of consciousness, can observe and control all mental activities on the lower levels.

It must be remembered that although there are different levels of consciousness, the mind is one, pervading all, but working through different levels of consciousness, its activity restricted by the limitations of each receding level or plane. That is what makes this idea so difficult to assimilate in the West—that a mind pervading and activating all levels of consciousness can on a high level rectify the working of itself on a lower level.

Yoga must be recognized for what it is and not for what so many think it is—a mysterious, unhallowed religion linked with extreme mortification of the flesh, sitting on nails, the Indian rope-trick, fire-eating, and bloody sacrifices at midnight in the jungle. Yoga is a system devised and practised for over two thousand years, with a therapeutic basis for its physical and mental objective. Much of it has been disguised by the symbolism so beloved by the East, but stripped of this it stands revealed as a rational system of culture, both physical and mental, and as such warrants attention by the Western world. The writing of this book is intended to bring to some sections of the Western peoples a realization of Yoga's great value to the individual, and, through the individual, to the nations.

The extraordinary powers of mind and body have long

compelled recognition by scientists, and in many cases these powers, not being understood, have been condemned as clever trickery. Fortunately, to-day, as never before, the East and West are meeting, and earnest workers on both sides are striving for a common ground on which to develop research for mutual advantage. Yoga is making slow but steady progress into the minds of Western men of science, and is providing adequate answers to many of the marvels and miracles of the East.

The yogi sees in man the conjunction of the gross world with the world of supreme power, knowledge, and Divine Essence that lies beyond. Man is for ever seeking his true position as a co-ordinate between these two spheres, and a knowledge of self is the realization of the co-ordination between the gross and spiritual planes. This knowledge is attained with the greatest speed, under the prevailing conditions of living, by means of Yoga.

Yoga is not a religion, but it advocates the acceptance of some divine source in which to bathe the ailing mind and to stimulate the necessary element in the student of adoration for the Divine. Yoga does not ask the student to prepare a scheme of supplication to a deity, for it is felt that then the independence of the individual is lost and a precious gift of God thrown away. Rather does Yoga reveal the God in man and put the responsibility of spiritual awakening on the individual, who must seek Him for himself by a practical and scientifically worked out method. The schoolboy never progresses when he continually asks his father to do his homework, and it is of far more value to that boy to work out one simple sum himself than have his father work out six.

Yoga asks for no blind belief, but offers an accumulation of attested facts for observation—not of external things but of the things of the mind from which true experience is born. It calls for a concentration in a healthy body of the mind on itself, and seeks to destroy desires, the seeking after which causes the striving and unhappiness of the world. With the mind's light concentrated, instead of diffused over a variety of irrelevant and oft-times useless desires, the bright light focused and turned inward reveals the true Self, and gives the yogi a destined path of true living and a calm clear outlook on all things, and that much sought after adaptation to life.

Yoga asks for no pursuit of extremes, but a logically worked-out middle course, which, with application and a fervent desire for knowledge, it is possible for all men to tread, irrespective of race or creed. Yoga carried out according to the rules, with sincere and sustained effort, brings physical and mental repose, with a renewed sense of optimism, efficiency, and self-control. It calls for no abnegation of the things of life, as so many tend to believe, and looks upon the indulgences of smoking and drinking with a tolerant eye, considering that progress in Yoga automatically destroys the need of what after all are drugs occasioned by nervous instability.

Yoga is common sense applied by system to meet the

bodily and mental ills of mankind, and as such earns the serious consideration of every intelligent individual.

PHYSICAL CULTURE, WITH EXERCISES

Chapter III

PHYSICAL CULTURE, WITH EXERCISES

Before starting to practise any of the exercises given here there is one thing of great importance that must be remembered by a student of the ancient yogic system of physical training, and that is that the correct attitude of mind must be brought to bear on this system if appreciable headway is to be made.

The student should have the utmost confidence in what he is doing, and should realize fully that he is an exponent of the world's finest culture for maintaining a healthy spine and abdominal area—for toning up the abdominal muscles, lumbars, and buttocks. It is a culture that demands poise and carriage, which are essential for healthy nerves, the mainspring of a harmoniously working body. Sitting in chairs has relaxed the abdominal muscles, which demand attention if their strength and elasticity is to be regained.

Know this, that the exercises here given cater for every need of the body, and that systematically practised and made a habit, a great improvement in the constitution of the student will result, making for a renewed outlook on life generally, of optimism, confidence, and happiness. These things will come if the student plays his part well.

Not more than three or four of the exercises should be practised at first, at any one time, but when proficiency is gained in all they may be practised in a series, systematically.

The Inverted Posture, although given later in the book, may be taken first when the student has mastered them all.

2. Preparation for Halāsana: the Plough Pose

Halāsana: The Plough Pose

The name of this pose is taken from its likeness to the Indian plough.

The student lies flat on his back. The legs are extended, the toes pointing up, and the arms lying straight at the sides, hands palm downwards.

(The exercise is sectioned off with slight variations as to posture, having marked differences on the regions of pressure.)

SECTION 1. The legs, kept straight, knees never being allowed to bend throughout active exercise, are raised from the hips 30 degrees from the ground, and held in that position for five seconds, influencing the contracting power of the abdominal muscles by reason of the lumbars being pressed against the floor.

SECTION 2. The legs are raised again until the angle to the ground is 60 degrees, and held for five seconds.

SECTION 3. The legs are raised now so that they point straight up in a vertical position. There is another short pause. It is important that the entire trunk should be pressed to the floor when raising the limbs.

SECTION 4. The legs are swung slowly over the head, only the hips and lower spine bending, until the toes easily touch the ground beyond the head. The toes are kept as near as possible to the head, without effort, with the legs straight, the toes pressing firmly on the ground. The hands and arms are still extended. Pressure will be felt now on the lower part of the spine.

In the next stage of this exercise the toes should reach

farther and farther from the head, the strain coming on the upper part of the spine. This should only be done slowly, and with care, to avoid overstraining.

For the completion of this exercise the hands and arms dissociate themselves from the levering process which swings the legs over the head. They are placed with locked or intertwined fingers against the back of the head, forcing pressure on the upper section of the spine near the neck. As the hands go behind the head the toes, which have been gradually sliding away from the head, are drawn in again towards the head.

This movement of the toes is very slight, but the student will readily note the different areas of the spine affected by each position of the toes when in the Plough position. At the close of the exercise the student reverses the posture steadily to regain the position he had prior to the commencement of the exercise.

The beginner should do this exercise three times, but a proficient student, after systematic practice, ten times.

As may readily be understood, this exercise done smoothly is of great benefit to the spine, and an elastic spine means a young body at any age, the lumbars also receive attention, the abdominal muscles are strengthened, and the neck in every way benefits.

The exercise is also conducive to a correct standing posture and to reducing the surplus weight of the body.

It may alternate with the Posterior-stretching Pose for increased beneficial results.

3. Halāsana: the Plough Pose

4. Pashimatāsana: Posterior-stretching Pose

Pashimatāsana:
Posterior-stretching Pose

The student sits on the floor, fully extending his legs in front of him. He then bends forward to catch hold of the right big toe with the right hand, and the left big toe with the left hand. The student then pulls on the toes, stretching the muscles at the back of the legs.

Next, he bends farther forward, until the head rests on the knees, the hands still holding the toes, and, if possible, the elbows resting on the ground outside the legs.

The exercise is similar to that common exercise of the West called 'touching the toes', except that in that the student stands up and bends his trunk down for the hands to touch the toes. There is, however, one vital thing that the two exercises have in common—the legs must be kept straight. If a student is stiff in the knee-muscles, their elasticity being poor, he should do this exercise grasping the ankles, or even the legs below the knees if the muscles are very stiff. The important thing is to keep the legs straight. Practice will bring the elasticity required for the complete posture.

It is sufficient to give three minutes to this exercise, which benefits the ham-string muscles behind the knees, the abdominal muscles, and the nerves of the pelvic region.

Sarvangāsana: Whole-body Pose

The aim of this exercise, which affects the thyroid gland, is to rejuvenate the whole body.

The thyroid is a gland in the neck discharging a secretion into the blood-stream materially influencing the nerve system and the human body generally. If this gland holds up the secretion the body is thrown out of balance and various parts affected.

The student lies flat out on his back with his muscles relaxed and his arms at his sides.

The legs are then raised slowly, making an angle of 30 degrees with the ground. They are held there for three seconds, then raised another 30 degrees, held, and then raised again, until they are perpendicular with the floor.

The body is then raised, still keeping the legs straight, the weight being thrown on the arms. When this exercise is done well there is little weight on the arms and hands, the body balancing itself when the legs reach upward. The chest is against the chin, and to increase the pressure the arms are bent at the elbows and the hands push the back. The neck lies flat on the ground.

When doing this with other exercises, never give more than five minutes to its practice in the early stages of training; with proficiency ten minutes may be allotted.

This exercise promotes general health by its healthy reaction on the thyroid gland, and in addition it benefits the sex glands of both sexes, countering sexual disorders. For women, it is a good exercise for a displaced uterus. It

5. Sarvangāsana: Whole-body Pose

Sarvangāsana : Whole-body Pose

is also efficacious for those suffering from dyspepsia, constipation, and hernia.

It may be described as an all-round exercise for improving metabolism (the balancing and harmonious process of cell-destruction and cell-building in the body) and for developing the disease-resisting power of the blood.

Bhujangāsana: The Cobra Pose

The semblance of a snake with its head raised ready to strike gives this particularly fine health-giving pose its name.

The student lies straight on the floor, face downwards. The soles of his feet are turned up, and his elbows are bent, with the hands level with the chest and the palms downwards.

The exercise starts with the head being lifted and the neck stretched upward. At this stage the chest is still kept as close to the ground as possible, with the trunk and out-stretched legs motionless and the toes in contact with the floor. The legs are kept together throughout, and not splayed.

As the chest begins to rise from the floor the pressure of support is felt by the hands, but proficiency in this exercise throws less weight on the hands and arms, and the tension is felt in the back muscles.

The gradual bending of the back, with the rest of the lower part of the body still, involves the raising of the vertebrae (the sectional pieces of the spine) one by one, and the contraction of the buttocks.

The student should feel the pressure on the spine work down section by section as he exerts it by raising the fore-part of the body up and backwards.

At first the bend achieved will throw pressure on the thoracic section of the spine. As proficiency is gained the pressure is extended to the lumbar region. An adept at the

6. Bhujaṅgāsana: the Cobra Pose

7. Preparation for Salabhāsana: the Locust Pose

Bhujangāsana : The Cobra Pose

exercise can bend the back till pressure is experienced at the sacrum region at the base of the spine.

When the student feels the pressure there he can reckon on having achieved the maximum results from this exercise. It must be remembered that all this time the legs have been outstretched and still.

The pressure should be withdrawn and the chest lowered to the floor as at the beginning, just as slowly and steadily as it was exerted. Allow the breath to flow evenly during this exercise, which should be performed from three to six times.

The Cobra Pose gives a blood flush to the spine, the sympathetic nerves and cells, and all the muscles of the back, and the spine gains that elasticity so essential to a vigorous and healthy body. Intra-abdominal pressure is increased, the stomach muscles developed, and flatulence averted.

An important rule to be remembered is that if a person is not supple before taking this exercise, the whole process should be done slowly and cautiously, and some time allowed to elapse before it is done with the complete spine exercised. The thoracic pressure should be got without harmful strain before proceeding further.

It is an exercise that should not be neglected, for, in addition to the benefits stated, its contraction of the buttocks allows them to perform more adequately their function of keeping the body well poised.

Salabhāsana: Locust Pose

There is a certain kinship in the position of the student doing this exercise and that of the locust with its reared abdomen.

The student lies stretched out on his face with the soles of his feet pointing upwards and his hands clenched at the sides, knuckles touching the ground. The chin is also rested on the ground.

The legs are raised quickly from the ground and kept stiff, pressure being felt on the arms and the lower part of the trunk. This exercise may be done about half a dozen times with safety, but a point to be remembered is not to hold the breath and the posture too long—not until it becomes really uncomfortable.

After finishing the posture and the breath retention, breathing will be rather rapid, and the posture should not be repeated until the acceleration of breathing has diminished.

The Locust Pose is the only posture in the whole of Yoga which calls for such sudden effort in raising the limbs.

It is a pose which might be described as the reverse of the Cobra Pose, for whereas in the Cobra the lower extremities were passive, in this posture the upper part of the trunk remains still.

It is obvious that this exercise, done smoothly, has a good effect on the pelvis and abdomen. The muscles of the back are developed and the general circulation in the lower limbs improved.

8. Salabhāsana: the Locust Pose

9. Dhanurāsana: the Bow

Dhanurāsana: The Bow

This position resembles a bow when it is bent.

The student lies face downwards on the floor and bends his legs at the knees so that the portion of the leg composed of the calf and foot is doubled back over the buttocks, allowing the hands to grasp the ankles.

The student then forces the grasped legs back, with the result that the top portion of the trunk is pulled up. The whole effect is a curve made by the trunk and thighs, and pressure is felt at the abdomen.

After maintaining this position for five seconds, release the backward pull of the legs so that you can rest against the floor, then release the ankles and return to the prone position.

The movements to attain the position should be brisk but smooth, and the breath allowed to flow as usual.

When starting the exercise a better curve is got by spreading the knees, but, as practice increases the suppleness of the joints, the legs should be drawn in together.

This exercise is a fine corrective for central curvature of the spine, and cures the gassy condition of the stomach, while the abdominal wall and muscles are greatly strengthened.

Ardha-Matsyendrāsana: The Twist

The student sits with his legs out in front of him. The left leg is then bent at the knee and the foot brought into the crutch, the heel being on the perineum. The right leg is so placed that the foot rests on the floor outside the thigh of the bent leg.

The left arm goes outside the upraised right knee (that is, not between it and the body) and grasps the toes of the right leg, which are squarely on the floor. The right arm is placed across the back at the waistline, palm outwards.

The head should be turned to the side (in profile), twisting the neck away from the upraised knee.

It is a pose that can be followed best from the illustration. The twist, it is seen, comes on the body away from the locked legs and arms, and to exercise the other side of the body the position of the legs and arms is reversed.

The chest is kept erect and the position, for beginners, maintained for a few seconds only.

A healthy body, and one of vigour, must entail a healthy elastic spine, and that spine must be exercised in different ways, of which this is one—giving the spine a side-twist.

The sympathetic nervous system is toned up, as are the muscles of the shoulders. The abdominal muscles get massaged, and the exercise is very good for the kidneys, liver and spleen, and for constipation.

10. Ardha-Matsyendrāsana: the Twist

(a) Front view

(b) Back view

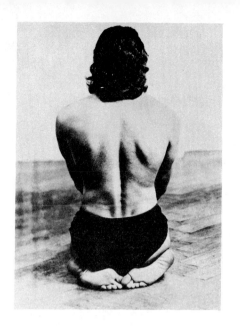

11. Supta-Vajrāsana
(a) Preparatory Position for the Pelvic Pose
(b) The Pelvic Pose

Supta-Vajrāsana: Pelvic Pose

The student kneels down, placing the feet with the soles uppermost, outside but close against the thighs. The back is kept upright and the posterior is firmly entrenched between the feet, so that one is almost sitting on them.

The student, in easy stages, lies on his back without moving the legs. The hands at first act as support for the novice, resting on the ground behind. The arms bend at the elbows, and the weight rests on the arms, and as the student bends back still farther the shoulders touch the floor. The arms and hands act as cushions folded behind the head when proficiency is gained. Alternatively the arms may be extended full length, palms up.

The student first finds his trunk bending at the waistline, but, with practice, he will make his back quite straight, as it is stretched back over the folded legs.

There should never be more than normal pressure on the back, arms, and above all, the ankle joints.

Retain the position about two minutes, and regain the normal position by reversing the process.

With this exercise the abdominal muscles are stretched, the pelvic region benefits, the stomach generally is stimulated, and the thighs are strengthened. It has also a very fine effect on those suffering from constipation.

Mayurāsana: The Peacock

The student kneels and brings his elbows together in front of his body between the knees, the hands being flat on the floor and pointing backwards towards the legs and feet, which extend behind.

The elbows are placed pressing into the stomach and, acting as a support and a fulcrum, or balancing-point, the body is raised and remains parallel to the floor, supported on the forearms. The position is like the letter T, with a short vertical line (the arms) and a long cross-piece (the body).

The head is raised and made to counterbalance the legs.

The position demands a great deal of energy.

By this exercise the blood is concentrated and sent to the digestive organs, which also get toned up by the intra-abdominal pressure.

It has a good effect on the expelling action of the colon, and is very beneficial for obesity, piles, constipation. It also develops the chest.

The whole exercise is regenerative and stimulative.

12. Mayurāsana
(*a*) Preparatory Position for the Peacock Pose
(*b*) The Peacock Pose

(a) Preparatory Position for the
Inverted Posture

(b) The Inverted Posture

13. Sirshāsana

Sirshāsana: Inverted Posture

The Inverted Posture is a posture in Hatha Yoga that will recall to many the physical feats of their childhood.

The student kneels on a thick mat, making sure the mat will not slide on the floor. A cushion should be used.

Kneeling forward, knees close to the body, he rests his elbows on the ground and brings his hands together, interlocking the fingers.

The back part of the head is then placed into the cup formed by the joined hands, and the trunk is raised up, the weight being on the elbows and the hands. When the student gets his sense of balance the legs are unbent and extended.

The student should do this slowly, as a sudden effort to extend the legs is apt to upset the balance.

If it is possible, get a friend to help you by holding your legs and informing you whether they and the spine are in a straight line. Get the friend to establish the balance for you, and try to hold it.

If you have no one to help you, kneel facing a wall, so that as your back and legs are raised the wall offers some support in case of losing equilibrium. Rest your upraised feet against the wall, then push off from the wall and establish your balance for about ten seconds. Try, however. to do the exercise without aid, and if you can hold your balance longer, do so, but make ten minutes a maximum.

This exercise should not be practised by people with bad ears, weak hearts, blood-pressure, nasal catarrh, or severe constipation, and never after any violent exercise. The precautions may be many, but the beneficial results are proportionately good.

There is a generous blood flushing from this exercise to the brain and every vital organ of the head, creating thereby increased efficiency for daily living. The glands around the heart also benefit, and the digestive powers are improved. The nervous system is also toned up, and altogether there is a remarkable improvement in general health.

As a pose it is invaluable for mental workers, for it sends an increased supply of blood to improve the functioning of the brain, pineal body, and pituitary gland.

An advanced student may adopt the inverted posture with foot-lock, shown in the next picture.

14. Advanced student's position in the Inverted Pose

Matsyāsana: The Fish

The basis of this exercise is the foot-lock described for Padmāsana (see page 68). To begin the posture, first sit in Padmāsana. Then, with the aid of the elbows, lie back, arching the back as much as possible, and raising the chest. The head may be fixed in position by supporting the body on the hands. When this has been done, the arms are extended and the forefingers catch hold of the toes on either side.

The Fish Pose is a marvellous exercise for the entire body, particularly the lumbrosacral region and the neck muscles. It has a reverse action to the Whole-Body Pose, and should immediately follow the Whole-Body Pose to attain the maximum benefits from both positions.

Yoga Mudrā: Symbol of Yoga

Yoga Mudrā Pose begins with Padmāsana, and in placing the feet in the hip-joints, the heels should press close against the abdomen. The hands are then brought behind the back, the right hand grasping the left wrist. The student should now bend forward, lying as flat as possible on the heels, and endeavour to touch the floor with his forehead. The movement forward should be slow and well controlled, and when the position has been held for two to five seconds the body may be raised to the erect posture.

Yoga Mudrā is an excellent exercise for the abdominal wall, internal organs, and the entire pelvic area. The pressure of the heels when bending forward is on the caecum and pelvic loop, with beneficial results to these parts in constipation. If the muscular wall of the abdomen is weak and the viscera have become displaced, the hands may be made to grasp the heels so that the knuckles press into the abdomen when the body bends forward. This action pushes the internal organs into their proper place and helps to relieve constipation.

The breath should move naturally, though exhalation at the time of bending is helpful, especially when the hands grasp the heels.

16. Yoga Mudrā: the Symbol of Yoga

17. Bandha Padmāsana: Meditative Pose of an Adept

Bandha Padmāsana: Meditative Pose of an Adept

For this exercise the student should adopt the Lotus Pose (see page 68) and then cross his arms behind his back and catch hold of his toes by the fingers of opposite hands. Those whose arms are rather long will be the first to succeed if they are not naturally stiff. This pose is good for the bony chest wall and promotes the expansion of the lungs to capacity. It is a meditative pose and has a beneficial effect on the whole body.

Simhāsana: The Lion Pose

The Lion Pose is a vigorous pose that imitates the fierceness of a lion about to spring. The student kneels, resting the buttocks on the heels and the extended hands on the knees. The chest is expanded and a sense of muscular tension pervades the entire body. The jaws are opened wide and the tongue thrust forward as far as possible. The eyes are widened in a fierce gaze, and the arms and fingers are stiffened.

The Lion Pose is invigorating and highly beneficial to the throat muscles, eyes, chest, and spine. The position may be retained for one minute in the beginning, and later on for three minutes.

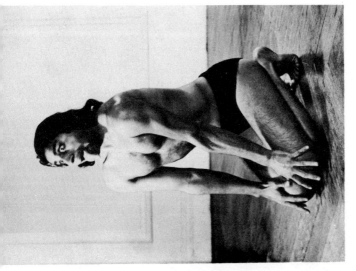

(a) Front view

(b) Side view

18. Simhāsana: the Lion Pose

19. Savāsana: the Supine or 'Dead' Pose of Relaxation

Savāsana: Dead Pose

The Dead Pose is a supine relaxing pose that offers complete rest and tranquillity to the body, mind, and nervous system. It should be the concluding pose for any group of exercises. It is simple enough to lie down flat on one's back, but systematic relaxation requires the aid of the mind. The student should fix his mind on the feet and relax them, then the limbs, arms, abdomen, chest, throat, facial muscles, and eyes should yield in turn to suggestion until complete relaxation of every part of the body is achieved.

An important part of relaxation is the breath flow. It will be noticed that the breath is uneven. Attention should be directed, not to increasing or lengthening respiration, but to bringing rhythm and evenness into the flow of breathing. This will help a great deal to secure complete relaxation, not only of the body but of the mind as well.

NOTE

Each of the postures described should be held when practising for two seconds to five seconds, and later on for ten seconds when the pose has become easier to manage. Every pose should be done only once at first. The individual must be the judge of his own capacity, keeping in mind that sudden and jerky movements will be injurious and ineffective. No more than three of the easier poses should be tried for the first week. The Plough Pose, Posterior-Stretching Pose, and the Whole-Body Pose form an effective trio of postures to start off with. Others should be added in the order given in this chapter.

PRĀNĀ—THE FORCE OF FORM WITH MEDITATIVE AND BREATHING POSTURES

Chapter IV

PRĀNĀ—THE FORCE OF FORM WITH MEDITATIVE AND BREATHING POSTURES

In this chapter the reader will have the opportunity of assimilating the great art of Prānāyāma—the different kinds of breathing exercises practised by yogis, the word coming from *prānā* (breath), and *āyāma* (pause).

It is essential, however, that *prānā* in its deeper sense should be understood, for in that understanding lies the true and full significance of Prānāyāma, or the science of controlled breathing.

Swami Vivekananda, in *Raja Yoga*, refers to that all-penetrating ether, Ākāsa, which he defines as 'the infinite, omnipresent manifesting power of this universe'.[1] The universal force that manifests what the Great Consciousness planned in the ether is Prānā, which for the West is made more comprehensible as the 'force of form'. Form, here, is not confined to length, breadth, and thickness, but to motion, gravitation, and, as Vivekananda further extends it, 'the actions of the body, as the nerve currents, as thought force. From thought down to the lowest

[1] *Raja Yoga*, p. 33.

61

physical force, everything is but the manifestation of Prānā'. This force of form is always existent, never extended or depleted, its existence only indicated by the extent of manifestation. Its control in its manifestation of man, is the science of Prānāyāma.

Its value is obvious, for if you wish for the manifestation in yourself of a fit body, or an increased awareness of the Great Consciousness, you can only achieve it through the controlled use of Prānā.

In attempting to control the Form Force (Prānā), we start with a great advantage, for we have two aspects of it, body and mind, manifested in ourselves. Investigating further, we find that there is one vital function which, if it is stopped, even though the body be in perfect health, brings death. That function is the motion of the lungs.

If you can control the motion of the lungs you can control the Form Force. It is not the air you breathe which works the lungs. The pumping action of the lungs is a manifestation of Prānā, or Form Force, so, if you control the pumping, you are controlling the Form Force that works them. If you hold your breath you are preventing the Form Force from deflating the lungs to expel the breath.

This, being the easiest approach to control of Prānā, is why breathing exercises take so prominent a part in yogic culture.

If one part of the body has an excess of Form Force, or a depletion of it, disease occurs, so the yogi aims at its even distribution throughout the body, and thus the feature of Prānāyāma breathing exercises is *rhythm*.

Controlling the Form Force in the body, and the whole body being a manifestation of this force, it follows that we control the motions of the mind, body, and especially the nerves, the gross channels for Prānic energy. Maintaining a balanced infusion of the Form Force over the body ensures psycho-physical co-ordination, when emotional complexes are relieved, and no longer does the individual become influenced by such unnatural products of an inharmoniously working mind as inferiority, envy, self-pity, jealousy, etc.

To some it may be difficult to appreciate how breathing, the mind, and the body can be so linked, and so we may give one very simple and ordinary example.

A singer goes on to the stage for the first time, an experience that frightens her and makes her nervous. Her legs tremble (physical body) because she is afraid (mind). The legs tremble because the emotion of fear has been sent by the brain to the legs via the nerves. She sings, and her voice is quavery and she cannot hold her top notes because she has not the *breath*. At rehearsals her breath came easily and she sang well, but now body, mind, nerves, and breath, are all working out of tune with each other and the singer fails.

The same lack of harmonious working of the complete man means failure in life. Before going on to the breathing in more detail I will quote another ancient authority,

Prāṇā—the Force of Form

Lu Tzu, who himself quotes from *The Book of Changes*: 'But even if a man lives in the power (air, Prāṇā), he does not see the power (air), just as fishes live in water but do not see the water. A man dies when he has no life-air. . . . Therefore the adepts have taught the people to hold fast to the primal, and to guard the One,' i.e. the Primary Cause.[1] This great Chinese sage describes in picturesque words how breathing comes from the mind, and the more fantastic the products of the mind, the quicker the breathing. He extols the opposite practice, for by breathing slowly and steadily stability of mind is ensured.

The harmonious regulation of breathing cannot be attained unless the body is in a correct posture. The ideal postures devised by yogic science are those which give the greatest freedom and poise to the body. The yogi sits on the floor with his limbs folded or crossing each other in various ways in order to perform breathing exercises and meditation. These postures, called Āsanas, give the pelvic region its natural position, allowing the abdominal muscles and the diaphragm to function properly in controlling the flow of breath. The upper part of the body easily assumes a well-balanced position over the hips, and correct breathing is then relatively simple.

There are many postures in Yoga advantageous to breath control and meditation, but as it is as well to achieve proficiency in them, or at any rate in one or two of them, the number given below is limited to four.

[1] From T'ai I Chin Hua Tsung Chih's *Secret of the Golden Flower*.

20. Sukhāsana: the Easy Pose

Sukhāsana: The Easy Pose

The Easy Pose is just as its name implies—a position of ease and comfort once the student can reconcile himself to doing without the upholstered 'comfort' of the West.

To attain the Easy Pose the student sits down on the floor with his legs stretched out in front of him. The right leg is bent at the knee and the foot placed under the thigh of the opposite leg. The left leg is then bent at the knee and the foot placed under the right leg. If preferred the position of the legs may be reversed.

The body should be kept straight with the head level, arms extended, and backs of hands touching the knees, thumb joined to first finger. This pose is the simplest one to achieve with conscientious application. It is excellent for prolonged studies in concentration and meditation.

If the student cannot at first tuck his legs in close, he must get them as near as possible without straining, and try every day to perfect his pose. As a help to those who in the initial stages are so stiff jointed that they cannot remain comfortable at all, sit on a book or anything that raises you two or three inches from the ground. It is surprising how this facilitates the disposal of the crossed legs. Of course, as proficiency is gained the book can be removed.

It will be found that the knees remain obstinately in the air, but with practice they will lower themselves to very near the ground. The knees should not be forced down; but if the space between them is widened as far as possible, and daily practice is kept up, proficiency will soon be attained.

Siddhāsana: Perfect Pose

One of the most popular poses for enhancing the faculty of concentration and meditation is the Perfect Pose. The limbs are first extended, then the left foot is grasped by the right hand and the left heel is placed against the perineum, just below the generative organs. The toes go under the right thigh, the sole of the foot being in contact with the thigh. The right leg is then bent at the knee and the right heel is placed against the pubic bone above the sex organs, the toes fitting into the cleft formed by the calf and the thigh of the left leg.

The left hand rests on the right heel, palm upwards, and the right hand is placed upon the left in the same position. The question of the hands is one that does not materially affect the results to be obtained so long as the beautiful symmetry of the position is not lost.

This is an ideal pose for concentration with the eyes closed, the chin pressed against the chest, the spine erect, and the gaze mentally directed to the space between the eyebrows in what is called the frontal gaze—a fine exercise for the wandering mind. This gaze may be practised with the eyes open, but a beginner is cautioned to restrict the exercise to short periods, starting from one minute and gradually increasing the time. The posture itself should be maintained according to the ability of the student, without causing a sense of strain.

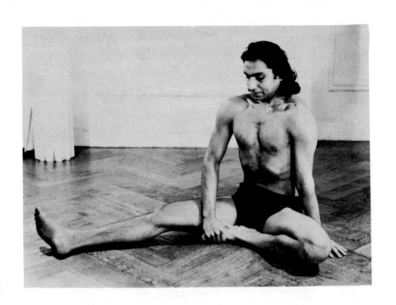

21. (*a*) Preparation for the Perfect Pose (placing the
heel against the perinaeum)
(*b*) Siddhāsana: the Perfect Pose

Vajrāsana: The Adamant Pose

The Adamant Pose is a simple posture which is included in the physical culture postures as an exercise for the pelvis. (See Preparatory Position for the Pelvic Pose, facing page 49.) The student kneels, keeping the knees close together, and rests the buttocks on the feet, which are placed with the heels turned out and toes meeting. The hands are placed upon the knees. As a variation, the feet may be placed soles up on either side of the buttocks. Flexibility will come with practice, as the pose itself is a great aid in preparing the body for the other positions.

Padmāsana: The Lotus Pose

The Lotus Pose is the most difficult of all, but it has cultural and therapeutic points in its favour which make it the most scientific posture for advanced yogic discipline, and it is extremely beneficial to the pelvic area.

The right leg is bent and the foot placed upon the left thigh close to the hip-joint, with the sole turned upward. The other foot is similarly placed on the right thigh, with the ankles crossing. The position of the hands is the same as in the Perfect Pose.

This wedge-like position of the feet is called the Foot Lock, and it enables the yogi to contract and manipulate the abdominal muscles in various exercises for health and spiritual progress.

It may cause considerable discomfort at first, but the results are worth the effort. Ability to perform the Lotus Pose varies with individuals. Some are able to take the position easily, others find it very difficult. It is a splendid position for the breathing exercises and for meditation.

22. Padmāsana: the Lotus Pose with Three Contractions
—Chin Lock (Jalandhara), Pelvic Lock (Uddiyana), and
Anus Lock (Mula Bandha)

NOTE

To many these positions or variations of 'squatting poses' will seem unnatural, but only because they have become used to Western habits. It is a good thing to make a habit of sitting each day in such a position on a rug for a while, to read a book or newspaper. It is amazing how eventually this position will always be preferred for true comfort. (It is well to try and increase the space between the knees as you practise 'squatting'.)

Older people may put all this from them because of inability to sit cross-legged, but when it is physically impossible to do so, the breathing exercises can be practised sitting on a low stool three to six inches below the height of an average chair seat.

When so sitting, keep the knees wide apart. Always sit erect and when breathing endeavour to draw up or contract the abdominal muscles, pressing the lumbars outwards and backwards. Do not raise or hunch the shoulders.

Simple breathing exercises may be practised lying down on the back, with the knees raised and the back pressing the floor at the waist. This gives no difficulty to those who cannot adopt the cross-legged positions. In fact, lying down is better in its ultimate effect than sitting on a low stool. It is not, however, the position to adopt for the advanced breathing exercises involving retention of the breath and contraction of the anal sphincters. Those who

69

are sensitive to the weakness of their abdominal muscles cannot do better than practise the physical culture positions of the Plough and Posterior-Stretching Poses, especially by raising the limbs up as in the preparation. It is important, however, that the student should do his utmost to adopt the cross-legged poses, for they are all possible to able-bodied people, with patience and practice, and are an essential part of the breathing exercises.

PRĀNĀYĀMA: SYSTEMATIC
BREATHING EXERCISES

Chapter V

PRĀNĀYĀMA: SYSTEMATIC BREATHING EXERCISES

The following breathing exercises are described in their logical sequence and should always be practised in this order.

Possibly the student will notice variations in the technique in several instances in some of the exercises. I do wish it to be understood that any deviation from other orthodox methods that there may be are following traditions in the main, and have proved practical advantages.

While we are accustomed to regard deep-breathing exercises as a means of oxygenating the blood, Yoga teaches us that it has a more important rôle. The exercises performed in the yogic manner are intended to activate the sympathetic nervous system and strengthen the intercostal muscles and diaphragm, which together constitute the respiratory mechanism.

By continuous practice of the exercises deep breathing becomes a habit. Also, these breathing exercises, by their action on the sympathetic nervous system, are the principle means by which Kundalini, or 'Key Energy', is aroused.

Prāṇāyāma: Systematic Breathing Exercises

The best time for the exercises is before breakfast. Nothing should be eaten, but a drink of water may be taken. It is most desirable to have a bowel movement before starting. If the exercises are done during the day, three to four hours should elapse after a full meal, or one and a half to two hours after light refreshments.

The exercises may also be repeated in the evening with advantage.

Kapalabhati

We start with Kapalabhati, as this preliminary breathing exercise is also a Shat Kriya—one of the six cleansing processes.

In this case it is the entire respiratory system and the sinus cavities in the head which are cleansed.

When doing this exercise, or in fact any of the breathing exercises, one of the meditative postures should always be used, preferably the Easy Pose for beginners.

Having taken up a firm and comfortable posture first exhale, and then take in a deep breath through the nose. Immediately after completing this inhalation, expel the air through the nose by an inward jerk of the abdomen. Involuntarily you will inhale again, when the jerking expulsion by the abdomen is repeated.

The expulsions are sudden and strong, and the whole process must be rhythmical.

At first the beginner should aim at one expulsion per second. He should try to do it five or six times in succession without a pause. Then he should take a rest and repeat the process. He should continue for three or four minutes, but with practice longer periods up to ten minutes may be attempted, and the speed increased up to two expulsions per second.

As the student becomes more proficient he may prolong the exercise up to thirty minutes.

The attention is entirely focused on maintaining *rhyth-*

mic diaphragmatic action, the speed of the exercise being a
secondary consideration.

It is a splendid exercise for singers, for whom the entire
course of Prānāyāma is highly beneficial.

Bhastrika: Bellows

Bhastrika, or 'Bellows', is so-called because the breath
current moves in and out vigorously with an audible
sound very much like that of the blacksmith's bellows.

Inhalation and exhalation are through one nostril only
—the other being closed with the fingers of the right hand
held in a specified manner called 'Vishnu Mudra'.

The thumb of the right hand is extended, the first and
second fingers are bent into the palm, while the third and
fourth fingers are straight. The thumb is used to close the
right nostril and the extended third and fourth fingers together close the left nostril.

To begin the exercise, press the thumb on the right
side of the nose and the two extended fingers on the left
side. Release the thumb and exhale through the right nostril, keeping the left closed. Then inhale suddenly and
quickly through the *same nostril*; close with the thumb
and allow the air to move out forcibly through the left
nostril. Inhale again through the left nostril, closing the
right, and exhale through the right, while closing the

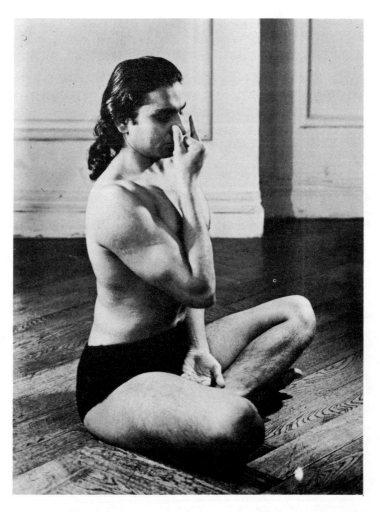

23. Bhastrika (Bellows): a breathing exercise

Bhastrika : Bellows

left. Exhalation is always followed by inhalation through
the *same nostril*.

Respiration is at first short and sudden, with a vigorous
impulse from the abdominal muscles in expelling the
breath. Breathing gradually becomes deeper, until a full
and deep breath may be drawn in evenly and retained for a
fixed period while both nostrils are closed with the fingers.
The Jalandhara Bandha, or 'Chin Lock', completes the
retention process, when the chin is pressed against the root
of the neck and exerts muscular pressure to close the throat
partially, and thus aid in the retention of the breath. (See
Lotus Pose with The Three Contractions, facing page 68.)

The proportional timing of the exercise is important.
If a full inspiration is made in eight counts, the breath
may be held for the same period, and then slowly let out
in twice the length of time (sixteen counts). The proportion of 1:2:2 is more beneficial once the student has made
headway: the breath should be held in twice as long as it
takes to inhale slowly. Advanced students may follow the
ratio 1:4:2, retaining the breath four times as long as the
inhalation period.

A beginner, however ambitious he may be, should *not*
attempt to retain the breath until several weeks of dili-
gent practice have passed in mastering the first stages of
the exercise.

There is no timing in the first part of Bhastrika. Timing is only necessary when deep inspiration is made and
the breath is slowly exhaled.

Prānāyāma: Systematic Breathing Exercises

When the student is efficient in Bhastrika and is able to retain the breath as directed, utilizing the chin lock, he may attempt the full technique of Mula Bandha, the Anal Contraction, as a part of the exercise (see page 85).

At the time of inhalation the circular muscles of the anus are contracted, and during the period of holding the breath, the contraction is intensified, being released only with exhalation, when the chin is also lifted.

The anal contraction should be undertaken with caution, as wrong method may lead to constipation and digestive disturbances. The anal contraction exerts an upward pull on these muscles and involves the upward contraction of the entire pelvis.

Bhastrika, as a breathing exercise, clears the nasal passages, increases the breath span to capacity, and purifies the entire respiratory system. It is an excellent method of arousing the internal vigour of the body.

In the preliminary stages of practice, Bhastrika should consist of ten or eleven expulsions, followed by a deep inhalation and slow exhalation. This series may be repeated twice, and after a few days' practice, a third round may be added, and deep inhalations increased to three each time. In advanced Prānāyāma daily practice may consist of more than sixty exhalations, with the addition of breath retention combined with the chin lock and anal contraction. The Perfect Pose, with heels pressing the perineum and pubic bone, is prescribed for advanced Bhastrika.

Ujjayi

Ujjayi is an exercise in slow, deep breathing. The air passing to and from the lungs is controlled by partial closure of the glottis, or windpipe. When taking a deep breath in the ordinary way, there is a frictional sound in the nose. In Ujjayi this involuntary frictional sound must be avoided.

The breath current moving in and out through a partially closed glottis causes a sound similar to sobbing, except that it is even and continuous. The best way to produce the sound is by using the imagination. For this purpose a mystic word, 'Hangsa', is thought of in the following way.

Having seated himself in one of the Yoga postures, preferably the Perfect Pose, the student first exhales completely, and then begins a slow inhalation, pronouncing *mentally* the first syllable of the word 'Hang' prolonging the sound for the duration of inspiration. Having completely filled his lungs, he then exhales with the same frictional sound, mentally pronouncing the last syllable 'sa' as long as the expiration lasts.

The abdominal muscles are kept slightly contracted during inhalation, and abdominal pressure is exerted when the breath flows out, and should be maintained until the breath is completely expelled.

Breath retention may be attempted after several weeks of practice, in the above manner. The frictional sound produced is a guide by which the student can judge the

Prānāyāma : Systematic Breathing Exercises

length and evenness of the breath flow. The complete process should be mentally timed to the ratio of 1:2:2 in the beginning and 1:4:2 later on when control is well established.

Ujjayi will probably be an easier exercise for the beginner than Bhastrika. It may be done immediately after Kapalabhati as a simple deep-breathing exercise, but once the Yoga method is understood the breathing exercises should be taken in the order here given. Kapalabhati activates the respiratory muscles and brings in rhythmic action. Bhastrika, besides clearing the nasal passages, gives depth to the breath span; and Ujjayi completely fills the body with vital breath, sending the breath stream into the sinus cavities in the head. *Action*, *depth*, and *expansion* are the three aims of the exercises in Prānāyāma in the order given.

Sitkari

Sitkari and Sitali are oral types of breathing with the lips open. The Easy Pose is a suitable position to begin with.

The breath is drawn in through the clenched teeth with the lips slightly open, and the tip of the tongue in contact with the teeth throughout the exercise. The lips are closed for exhalation, which is done through the nostrils. The breath stream is cool and invigorating, and is said to have a cooling effect on the system. During exhalation the breath current can be heard in the ears as it moves up into the sinus cavities.

Sitali

The tongue is extended well forward beyond the lips, and the two edges are rolled upward to form an aperture through which the breath is sucked in. The mouth is then closed keeping the tongue against the teeth as before, and exhalation is done through the nostrils. The breath current again moves up into the sinus cavities, and can be heard inside the head.

Both of these exercises produce audible sound, which is an essential part of the technique.

24. Uddiyana Bandha: Pelvic Contraction

Advanced Exercises

Uddiyana Bandha: Pelvic Contraction

'Bandha' means 'lock', involving muscular contraction of certain parts of the anatomy. Uddiyana is a muscular contraction in the pelvic region or lower abdomen which is utilized in expelling the breath.

The student, having taken up a firm seated position, makes the deepest possible exhalation, then raises the shoulders and bends forward slightly and expands the chest cage without inhaling.

The effect of this is to make the abdomen more and more concave until it appears almost to touch the backbone.

When it is felt that the breath cannot any longer be held out comfortably, the student begins slowly to inhale and relax.

Beginners should only attempt this exercise once a day and it must always be performed on an empty stomach.

Uddiyana is combined with Nauli, one of the six cleansing exercises of Yoga, whereby the internal parts of the body are cleansed or purified. Utilizing the technique of Uddiyana while the breath is held *outside* the body, the two recti muscles of the abdomen which extend from the ribs to the pubic bone are isolated together, and separately or alternately. A rotating action of the recti muscles is

83

then produced by alternate and rapid contraction of each rectus in turn, thus giving the entire abdominal region a rotary massage, with marvellous results to the health of the internal organs. Manipulation of the abdominal recti muscles gives the yogi complete control over the colon.

When the recti muscles are isolated, a partial vacuum, is created in the colon which enables the yogi to draw up water into the colon through the rectum. By means of the rolling manipulations of the recti muscles, a cleansing massage is then given to the colon canal, after which the water is automatically expelled.

Similarly, a partial vacuum is created in the bladder during this exercise, which enables the yogi to draw water into the bladder through the urethra. This remarkable suction power is called Vajroli.

25. Uddiyana Bandha: Pelvic Contraction
(a) Madhyama-Nauli (central Nauli)
(b) Uttar-Nauli (left side
Nauli)
(c) Dakshina-Nauli (right
side Nauli)

Mula Bandha: Anal Contraction

The anal sphincters are two circular muscles, one internal, and the other external, situated in the rectum. The anal contraction is practised as a part of Prānāyāma, or breath control. The vital energy of the body is drawn up from its centre in the Muladhara (sacro-coccigeal) region, when the breath is retained in the body for a fixed period.

Neti Kriya: Nasal Cleansing

Neti is one of the six processes of purifying the body. A soft cotton cord is used by the yogi to clean the nasal passages, which are thereby kept healthy. Catarrh and other diseased conditions of the nasal area are prevented by Neti Kriya. The olfactory nerves also benefit, thus improving the sense of smell.

26. Neti Kriya: Nasal Cleansing

27. Dhoti Kriya: Stomach Cleansing

Dhoti Kriya: Stomach Cleansing

Dhoti is a special process for cleaning the stomach. A long piece of wet cloth like a narrow bandage is swallowed slowly in its entire length. After which the Nauli action of rotating the recti muscles is used to give a cleansing massage to the stomach, which removes accumulated mucous. This exercise keeps the stomach healthy.

These two Kriyas should not be performed without proper supervision, and are referred to merely for the sake of completeness.

Shāmbhi Mudra: Concentration

Through this Mudra the nine doors of the senses, or the nine openings of the body, are closed. All external distractions of the mind are shut out, by placing the fingers over the two eyes, ears, nostrils, and the mouth. The urethra and the anus are effectively blocked by the pressure of the heels on the caecum and pubic bone.

This Mudra is one of the finest and simplest poses for concentration; and as such is recommended by Lord Shri Krishna in his famous work. (See illustration facing page 92.)

CONCENTRATION AND
MEDITATION

Chapter VI

CONCENTRATION AND MEDITATION

In a world where a diversity of things assails both body and mind on all sides, it is not unnatural that, to use an expressive Western phrase, people should find it difficult to see the wood for the trees. In the East it is an estimable virtue that enables the individual to project his mind on one subject to the exclusion of everything else.

In the West this is also recognized as a virtue, but as one so very difficult of attainment that the effort is never seriously made. Many of the great men of the West have achieved their greatness by singleness of purpose but they are the exceptions, rather than the rule and, because exceptions, have risen from the crowd to greatness. The East does not claim an inherent ability in this direction, and being at a distance, looks on with a detached and unruffled air at the Western world, so wrapped up in the practicalities of a materialistic age that the true blessings of life have been warped or lost sight of.

The Westerner finds it difficult to achieve singleness of

purpose because he has encumbered his horizon with diverse things that keep his attention in a state of continual abortive flux. The Easterner has not those distractions, and so with greater ease controls his mind.

Both East and West have contributed greatly to the advancement of mankind—pooled, how infinitely richer would these blessings flow. This book is but another effort in that direction; for the day will surely come when the knowledge of the East and the West will be linked, to the mutual advantage of both, and it is this unassailable confidence in the future that keeps pioneers at work in this field.

CONCENTRATION

As the body demands attention and an awakening to its potentialities, so does the mind, and Yoga meets this as it does all the real necessities of the individual, with a rational system proven by tests through hundreds of years.

When a person refers confidently to his ability to concentrate, it indicates no sure knowledge of the subject, for many making such a claim, refer to superficial efforts over a short period, whereas the process, simple as it is in regard to its technique, when practised more intensely over a longer period, would be found to be fraught with difficulty.

One of the most common pastimes in Europe is cycling, and it provides an excellent analogy for this subject, for the preliminary difficulties are much the same. At the

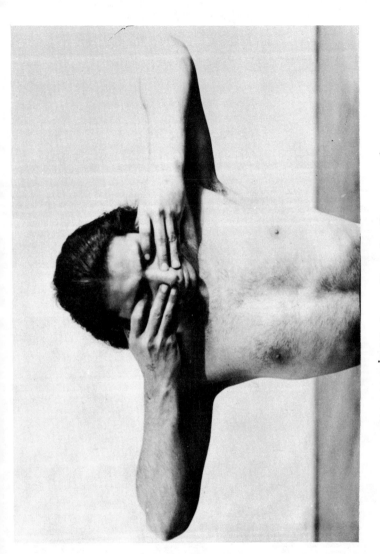

28. Shāmbhi Mudra: a Pose for Concentration

first attempt to ride a bicycle the learner is taken up with the fear of falling off, the mind flits from the feet on the pedals, to the steering, a super-human effort is made to absorb everything at once, control is lost, and the rider falls off on to the ground. But after constant application the learner finds himself riding easily, and the former desperate efforts at concentration on a variety of things concerning the bicycle appear in retrospect as amusing.

Ask an habitual rider of a bicycle if, when riding, he thinks of his saddle, his pedals, his front wheel, his steering, etc., when going along the main street, and he will answer in the negative. It is the same when teaching the mind to concentrate. If the student tries to hold it to the object by forceful methods and thinks at the same time of whether he is comfortable or not, and about the actual process of concentration, the object slips from the mind and is lost.

When concentrating on an object it is best to begin slowly and concentrate for thirty seconds intensely rather than three minutes in a half-controlled, disconnected fashion. Choose any object for the mind to fasten on, and make it a pleasant one—say a half-opened rose. Place it in front of you in a little vase on a level with your eyes, and think of nothing but its beauty, the soft imperceptible grading of the colours, the stalk, the thorns, and the leaf. Keep your mind on that rose until you could tell anyone every feature of the blossom, and eventually, even with your eyes closed, that rose will be so firmly imprinted on your mind that it will be as vivid for you as if you were still looking at it.

Regard the exercise as a kind of game, and have a shot at it at odd times of the day when you are waiting for a bus or for a meal. Do not waste that time by idle dreaming, but consciously direct your mind to some object, and, shortly, an admirable technique of mind-control will be established.

The student gives his mind a definite job to do, and what seems to the casual observer wasted time becomes to the student time of value and benefit. This process regularly applied prepares the mind for meditation.

Many may still consider this easy, but if you experiment and are honest with yourself you will admit that in the process alien thoughts will keep drifting across the main theme and disturbing you. Be very careful in your treatment of these thoughts, because as they appear and you try to force them back, their power is intensified, the law of reverse-effort is introduced, and concentration lost. Instead, take the thoughts as they come, admit them, and then deftly turn them aside, all the time keeping the main theme uppermost. It is a process whereby only practice brings success.

The ability to keep one thing in the mind irrespective of outside influences is called by various names, but one as good as any is uni-focal, maintaining the mind on one focal point. 'One-pointedness' of mind is another phrase much used to denote this fixity of the mind on an object.

Concentration

This restlessness of mind that we experience when trying to concentrate seems abnormal when the effort of concentration is being made, but actually this is not the case. Normally a restless mind carries with it a restless body, and the body, going with the mind to some extent, minimizes the awareness by the individual of how really restless his mind is.

If two trains were travelling in the same direction parallel to one another, although one might be going at fifty miles an hour and the other at sixty, we, travelling in the one, could not appreciate the speed of the other unless we were stationary. It is the same with concentration and the mind. When the body is still, we become aware of the true speed of our thoughts, while the speed, in effect only, is diminished if the body is also travelling, as it does in the normal routine of living.

The finest yogic practice for bringing the wandering mind to heel is the repetition in sonorous fashion of the resonant syllable 'OM', a sound which, starting in the throat and ending on the 'M' sound with the lips, includes all the lip and throat movements in speech, and so has a special significance in that respect as a totality of sound.

The symbol of OM painted green, about four inches in height, is also a traditional object on which to concentrate, and, as a beginner, the reader need not take into consideration the full significance of the sound or symbol, but at the same time, its value as an object for concentration has proved for years unassailable.

When the mind begins to wander keep saying aloud without a pause, 'OM,OM,OM', until the object reasserts itself in the mind. It can also be said mentally with effect when the student has progressed a little.

After some little research into the mind it begins to dawn on us that it seems to do anything but stand still. Like quicksilver on a moving surface it flows here and there, defying all efforts to check its erratic course. To adopt another analogy, it is seemingly as free as the wind, coming from we cannot tell where, and fading away before we know whether it has reached its journey's end or not.

It is essential that this wandering vagabond should be halted. Of the processes given to attain this end all should have an essential prelude—the physical fitness of the body, for the harmonious man must be healthy in body for complete success in achieving a healthy mind. If the body is not strong and healthy it is impossible for the mind to work efficiently.

Physical cleanliness is imperative, and that means that, besides an outside soap-and-water cleanliness, the inside, the bowels themselves, must be clean, until, altogether, one must have the joyous fragrance of a purified body, an intensification of that light airy feeling of well-being experienced after a good hot bath.

With this initial cleanliness achieved and made a habit, as it should be, we are in a better position to approach methods of harnessing the mind to go in the direction we want, and not the direction or lack of direction the mind has such a passion for. Exercises in concentration are fruitless unless the mind learns to fix itself on one point, and to hold it, oblivious to all the things that pull in contrary directions.

When putting concentration to practical application, decide to do something (a right action of course) before the day is through. Make it something that is possible to be done in a day or part of the day, and, allowing nothing to stand in the way, do that thing. That is what might be described as active application, but an instance of passive application has been given in the previous exercise, and though more simple in procedure, presents more difficulties. The student can easily invent exercises for passive concentration for himself, and as a further aid to carrying out the general plan for learning concentration, two more exercises are here given.

Place an acorn before you and gaze steadily at it, fixing it firmly in your mind. Know with certainty its attributes, as with the rose—its shape, colour, and any markings you can discern. Then imagine the acorn in the ground and a root shooting from its burst seed, and the first green leaf, and carry on the process to the complete tree. Leave nothing out, hurry nothing, let nothing interfere with you until in your mind the acorn has become a great tree.

A great aid to concentration is to close the eyes and direct the gaze mentally between the eyebrows, and to keep it there, allowing no idea or thought to intrude.

Concentration and Meditation

Breathe as quietly as possible, then you will become acutely conscious of the fact and of the beating of your heart. A dog may bark, a motor-car pass by, the ticking of a clock may intrude on your attention, but try mentally to wrap yourself up in a blanket of silence until all sounds fade, consciousness of the body vanishes, and a sweet sense of complete repose steals over you. Try for five minutes daily, and then later increase it to ten.

Another attractive subject for meditation is to visualize a candle burning steadily in a dark room. Some advocate fixing the eyes on a fire in the room, which many will admit has an attractive soothing effect conducive to the absorption of the mind, but its physiological effect on the eyes makes it inadvisable.

An adequate power of concentration allows the student to give undivided attention to an object or an abstract thought. The sure, full, knowledge thus gained allows for more competent judging of values. Before concentrating on the abstract, however, one must be proficient with visual objects, and their absorption and retention in the mind.

When concentrating, do it easily, and without such effort that the effort itself distracts the mind from the object contemplated. The command of the mind gained from concentration brings a mental repose and a new confidence which is reflected in all the dealings of the individual with the outside world. Increased efficiency gives the individual a better position in the world, and everything done is done well because no mental energy is wasted or uselessly frittered away.

MEDITATION

Meditation is the natural development in yogic training of concentration. Whereas concentration trains the mind to absorb the mind's objective exclusive of anything else, meditation is the process whereby the uni-focal mind is thrown on to great spiritual truths, and might be said to impinge on that part of developed concentration which deals with purely abstract ideas.

It may readily be appreciated that an abstract thought would be valueless and even dangerous without the preliminary training of concentration and the mental power of fixation of a mental object. The focus of the concentrative mind is on a small area of mental vision, with the consequent result of revealing that area in startling clarity of detail. In meditation that focus is extended over an infinitely larger area, but the mental light is maintained in its brilliance, so that the larger area is as conspicuous with its detail as was the limited area of concentration.

The stages of meditation can be roughly estimated by the student's individual power of extending the area upon which the mental light is focused without one iota of clarity being lost. Meditation takes into its ken such objectives as love, sacrifice, service, beauty, loyalty, the flow of life, the impermanency of matter, and, in fact all those great abstractions and virtues which are described as the true realities.

When the mind is educated to dwell beneficially upon these great things, and is so disciplined that all the forces of distraction are shut out, meditation is the method whereby the mind is kept facile and steady and in an harmonious contact with life, which creates the blessing of true inward happiness under any external circumstances.

The tranquillity of a mind composed is a great thing in a world of rush and bustle and nervous strain. It allows strength to be fostered, which can be exploited, when the time is ripe, in concentrated form in a single direction to make a man great in the eyes of his fellows.

This matured, well-trained mind begets a feeling of content that one cannot appreciate until it is attained, and with a healthy, harmoniously working body the individual becomes a king of the most important country in the world—himself.

Meditation has a positive and negative pole, or, in other words, two extremes in its practice which must be avoided.

A person who erring on the positive side, avoids the great middle path, develops an assertive inner nature, and takes pride in a false superiority of power.

The student who sees no difference in receptivity and surrender of the will meditates in a negative manner, laying himself open to every passing force, desirable or not, instead of practising disciplined reception.

It is not necessary in a book of this nature to deal with

the effects of extended meditation in detail, but it is advisable to conclude the subject with a warning in respect of self-hypnotism and auto-intoxication when ecstasy transcending all earthly bliss is experienced, and a phantasmagoria borne on a super-sensual plane plays on the student's mental horizon. Accepted for what it is, this ecstasy cannot prove harmful, but taken as the goal itself it stunts mental growth at its most promising stage, acting like the keen wind that nips a blossom in the bud.

This state can be brought on by varying means, and is that condition in which certain people see visions and hear the voices of angels, and become inspired to found sects and cults the world over. All this need not worry the average reader, however, for it is sufficient to conduct oneself in daily life as a respectable and honest citizen, reflecting that calmness and contentment of mind gained from the practice of meditation and concentration. It is not necessary to allow a sense of humour to be destroyed in the path of mental training, for to laugh at a thing denotes an ability to see it in perspective, a quality often much lacking in the world. Progress in mental exercises of concentration and meditation gives a new poise to the mind, reveals its hidden potentialities, and allows it to assess values more truly.

The final stages of meditation so refine the working processes of the mind that things become known and appreciated by intuitive knowledge, that neither logic nor scientific reasoning can understand.

Concentration and Meditation

At its supreme stage the student meditates with the objective part of himself, while the meditator and meditation are one, each identified with the other.

Meditation is an experience bound by no rules, and in describing it words become inadequate, but with physical culture and breath control it is one of the primary aspects of all Yoga.

29. Symbolic Meditative Posture

KEY ENERGY

Chapter VII

KEY ENERGY

Hatha Yoga, from the technical standpoint, provides one of the finest systems of physical culture known, and in the main it is on this side that emphasis in this book has so far been put, yet it would not be a proper completion of the task of expounding Hatha Yoga to ignore or gloss over as a mere superficiality what is the great end of this culture—the arousing of Key Energy, or as the Indian symbolically describes it, the Serpent Power of Kundalini.

Hatha Yoga has this to recommend it, that it is a system of culture that may be practised beneficially by those desirous of attaining physical fitness and mental alertness, and also by those wishing to know something of the spiritual experiences attendant upon specific exercises that come under the system.

I am well aware that many readers, perhaps the vast majority, are interested only in the physical aspect of this culture; to them is given perhaps the finest, and probably the oldest, systematized course in the world for attaining their purpose.

Key Energy

Yoga is a vast subject, incorporating a great many profound aspects of life, all of which interpenetrate, but which can be considered severally in the initial stages of practice. To those who wish to go a step further, when the healthy mind in the healthy body wishes to assimilate some of the truths surrounding its existence, this chapter is dedicated.

In no way does this affect what has already been said, but, to those wishing to penetrate a little further into the more subtle depths of Hatha Yoga, it provides an introduction to a little-understood subject of vast import.

It is postulated that throughout the universe is Consciousness, which we may well call God. It is better to refer to 'Consciousness', as there is always a tendency to build up an anthropomorphic idea of God as a venerable old man of infinite power and wisdom, which, instead of suggesting God symbolically, merely indicates the limitations of the average person's powers of conception.

This great Consciousness is Power in two aspects—static and active. The active sends forth its vibrations (or influence) and creates form for specific manifestations, and yet, in so doing, it hides itself from our knowledge behind the form it creates; as a wall built by a man to hide the sight of pedestrians walking by also has the effect of preventing the pedestrians from seeing the maker of the wall.

The instrument through which creative Consciousness

works must be something that is in contact with all things if it is to have an effect on all things.

Physics have now reached a stage when an intangible medium is postulated through which light travels, sound vibrates, and electricity operates, through which work the forces of attraction, repulsion, and cohesion, in which the atomic structures of all things float and have their being. Western scientists call this substance (or lack of substance) ether, while the Indians call it *Akāsa*. We can readily see that if this ether is an instrument of creative Consciousness, and we, as human beings, have the ether intermingled in our whole body, every atom of which is separated from the next by the invisible ether, then that which works through the ether must work through us as indeed it works through everything that exists. By this medium the forces are at work which hold all the atomic structure of the world together in a harmonious whole.

The form created obscures for us the realization of that great Consciousness that, in and outside us at the same time, is responsible for every activity of thought, deed, and reaction. We think it is matter that acts, and not the activity of Consciousness intermingled and working through it, which is the true faculty.

For instance, in one of the books referred to later in this chapter it is illustrated how through hypnosis a person is made to smell not through the nose but the toes. We think it is the nose that smells, but it is because we pay too much attention to the instrument, form, through which Consciousness works. It is not the form, the nose, smelling, but the faculty activated by Consciousness through form.

Matter is nothing more nor less than the result of vibrations of the atomic structure activated by the energy of the great Consciousness.

In the East it is believed that matter is in effect an illusion, and this explains in great part the reason why the East, compared to the West, attaches relatively little significance to the things of the world, which for Eastern minds are but illusions conglomerated into one great illusion.

Man has within him that vast power of Consciousness, but only in its passive aspect. The great energy which could work on the sensitive governing centres of the body, increasing their output, is nearly at rest, yet the ether-medium is there, ready for it to flow through and make its re-invigorating contacts.

If man aroused that great power of Consciousness called by the yogis 'Shakti', which is part of the Universal Consciousness, he would have, if he realized its maximum potentiality, that great illumination which Christians most happily call the Light, and so be in a position to yoke himself up (Yoga) with the Divine. To the Indians this great power of Shakti, known to the Western world as inspiration, or the divine spark, is a name covering one of the greatest forces of the universe.

Key Energy

Now if the form obscures the realization of the nature and power of the Consciousness which created it, we must obliterate the form as far as possible, and to the degree by which we destroy the obscurity of form, to such a degree is the awareness of the all-pervading power of Consciousness made evident.

That is the great object of Hatha Yoga. Create a perfect body, and as its creator, its lord and master, hold it in discipline and reduce the manifestations of its existence in the world of matter to a minimum. The same thing applies to the mind (which is only a more subtle instrument of the great Consciousness), hence the mental training in concentration and meditation already touched upon.

Arthur Avalon in *The Serpent Power*, addressed to students of the West, is responsible for a theory by which the static power of Key Energy, or Kundalini, is the 'coiled spring' (coiled, sleeping serpent) that lies at the base of the spine, and by the process of special exercises there is a reaction whereby the passive becomes active and rushes up through the body like an inductive current (discharging power, but not losing that from its source) on an unseen magnetic course (called Nadi by the yogis), re-vitalizing sensitive centres of the body corresponding in position to the plexuses, but not to be confused with them, as they are but the grosser manifestations of invisible and more sensitive centres.

The goal of this force is the brain, and if maintained there by the will of the yogi great and indescribable bliss results from an awareness of the great Consciousness that is God and which is in man.

This Key Energy may strike many as so intangible an idea as to be put from serious consideration, but nevertheless it is worth considerable thought.

Key Energy is an apt description of this force latent in man because once aroused and directed, like a key, it unlocks the many closed doors of the mind and the important centres of the body. Like wireless waves, it can be utilized and directed, and certain results accrue, but no man of science can dogmatically assert the true nature of such results.

V. G. Rele, in *The Mysterious Kundalini*, examines Avalon's theory more closely. It must be borne in mind, however, that nothing has ever been satisfactorily established as to the exact nature of this Kundalini force, and that everything said about it is vague conjecture, but this lack of definite knowledge is no indication that the cultivation of this force should be abandoned. We do not as yet know, as we have already said, the exact nature of wireless waves, but that would be the last reason a person would give for denying himself the pleasures of the radio.

The body of a man contains within itself, and as part of itself, a certain mechanism that enables him to realize that he is in a cold room sitting on a hard chair, smelling the paint on the walls and seeing its colour as blue, while behind the walls he can hear mice squeaking. In other words, this mechanism allows the man to be conscious of

110

his environment, and to react accordingly. This mechanism is the nervous system of the body.

The nervous system may be described as a network of telephone wires going all over the body, linking up every part. They have to pass their messages from all parts of the body to the brain, which is the great centre of the nervous system, and has for its assistant the spinal cord, which sends out some of the wires from itself to different parts, takes the messages, and then hands them on to the brain, as well, of course, as receiving messages from the brain and distributing them to the requisite quarters.

Among the nerves that leave the brain are two great cables called the vagus or pneumogastric nerves, starting from that part of the brain called the bulb and linking the brain to the vital organs, the heart, lungs, stomach, and intestines. The most active of these cables is the right one. It is a cable running parallel to and outside the spinal column, and it works, according to prevailing knowledge, automatically. It is part of what is described as the parasympathetic system.

To illustrate its workings, imagine a person seeing someone he has understood to be dead. The brain sends an impulse of extreme surprise and overpowering joy down the vagus nerve; at the requisite level the impulse branches off from the spinal cord into the cardiac plexus (the nerve centre of the heart), and, in sympathy with the brain's message, the heart restricts its action and the person faints.

111

Key Energy

Medical science says that this whole process is automatic, and one over which we have no control, or at any rate no conscious control. Hatha Yoga, on the other hand, teaches the establishment of control over this usually automatic action by a use of Yoga.

To carry our example still further, it is obvious that if the person, despite the shock of surprise and great joy, had control over the vagus nerve energy, that person could have arrested the impulse before it restricted the heart action and caused the faint.

That is only one example of the value of control over the para-sympathetic nervous system. Investigated in its most profound and spiritually important aspect, we can see that such control allows for a more intensive effort towards spiritual growth.

Every activity of mind and body demands energy. If a great deal of energy goes to one particular place, that place will be proportionately the more active. It would be impossible for the mind to be exercised to an unusual degree in an effort at spiritual discernment if there were a lack of this plus-energy to the mind. In the average case there *is* a lack of plus energy, because energy is being taken to supply motive power to the abdominal regions, which, being given some irritant and functioning irregularly, demand a countering process to restore equilibrium. The same divergence of energy might apply to any other part of the body, and for a variety of reasons.

The yogi aims at keeping his vital organs in a healthy

state, so that they need a minimum of energy, and instead of letting the energy lie dormant he whips it into activity by the stimulating effect of breathing exercises, and consciously directs it to where he desires it to contribute a plus amount of energy. Ultimately, with the spiritual objective in view (the objective of all Yoga), the energy is directed to the mind, and with its plus-quality actuates the mind to discern things spiritually to a plus-degree.

In his book Rele describes why the special breathing exercises play so large a part in the arousing of Key Energy. The respiratory act is under the control of the vagus nerve, the fibres of which are excited by alternate contraction and distention of the air vesicles at the termination of the nerve.

The vagus nerve is part of the sympathetic system, but being in function antagonistic to the reactions of the sympathetic, is termed para-sympathetic, contributing the anabolic or energy-conservation function to the body, of which the katabolic or energy-exploitation function is the antithesis. The vagus is vital in that fibres of nerves go to the heart, bronchia, gullet, stomach, intestines, pancreas, and also affect the rectum, anus, and genital organs.

The seat of the power which is projected in a direction corresponding to the vagus is given as at the sacrum. The symbolical language of the Indian yogi seems to indicate that the power lies semi-dormant in the prevertebral plexus of the para-sympathetic system.

Key Energy

It is outside the scope of this book to enlarge on the intricate details of the nervous system. It is sufficient to say that an interesting theory has been advanced to indicate that the Kundalini power is the activity of the right vagus nerve, or pneumogastric nerve, as it is called, and that the Chakras, or vital centres, revitalized by the energy aroused are in fact the nerve plexuses, the Chakras given by the East roughly corresponding to the position of the plexuses. The whole theory cuts across accepted medical knowledge and pre-supposes additional potentialities to these plexuses and the whole of the nervous system, which, it must be admitted by every medical man, is not completely charted.

That facet of Yoga which deals with the arousing of this Key Energy is held in high esteem by yogis, in that the realization of the oneness of the consciousness in man with the Great Consciousness, apparently outside him, comes from a united effort of both body and mind, and these are both shown to be aids, not barriers to the apprehension of things divine.

ABOUT FOOD

Chapter VIII

ABOUT FOOD

In ancient India, about 200 A.D., there lived a physician known as Charok, who was attached to the Court of King Kanich Ka. He compiled a work called *Charaa Samita*, which is a very comprehensive treatise on the cause and cure of disease. Even in those days when food was plentiful and could be picked and eaten in a fresh state, without the vitamin loss attendant on modern marketing processes, he stressed the value of a good diet for maintaining a healthy body and a well-balanced mind. To quote directly from this sage, 'One should take a proper measure of food,' this depending upon the power of one's digestive fire.

He classified the articles of diet as heavy and light. To the light articles he attributed the properties of air and heat, and to the heavy items the properties of earth and the moon. He therefore advocated that the major part of diet consist of the light articles because of their capacity to enhance the digestive fire, such food as fruits, vegetables, whole rice, fresh milk, cheese, and butter. He stated that the heavy foods consisting of meat, fish, eggs, poultry, wine, and so on, should be used sparingly, because they were injurious when used to the point of gratification. His teachings emphasized that the right food would promote health, strength, and longevity. It is interesting to note in view of our modern use of liquid diets, fruit and vegetable juices, that this sage said, 'Liquid food alleviates hunger, thirst, fatigue, weakness and disease of the stomach.'

Naturally, those interested in the Yogi way of life want to know what a Yogi eats. First of all he is a vegetarian, living principally on a diet of raw and cooked vegetables, fresh fruits, nuts, rice and fresh milk, also oil and ghee, the ghee being a clarified butter. He also uses an unleavened bread made of freshly ground whole wheat flour. This is made into a dough and then broken up into little flat cakes before cooking.

There is much for us to learn from this Yogi approach to food needs. Modern dietetic teachings emphasize the necessity of at least some raw foods daily, for the most revitalizing of all are raw food meals. In our fruits, vegetables, nuts, and seeds are found all of the vitamins, minerals, amino acids and proteins, unchanged by heat, and therefore immediately available for the body's replenishment needs. Beautiful and tasty salads can be prepared from luscious leafy green vegetables, blended with nature's own bright and gay colours. Beets, carrots, tomatoes, various succulent herbs such as watercress, parsley, radishes, chives, can be mixed. Three or four of these with a little simple dressing, a little grated fresh coconut, a few raisins, and so on, eaten with some nuts or cottage cheese, make a wonderful and tasty meal.

For hot dishes there is as much variety as one's imagination can conjure up. There are many types of beans—soya, mung, lima, navy, garbanzo, and so on—which, if cooked with a little onion and perhaps garlic, with a pinch of herbs such as marjoram, sweet basil, thyme, and so on, can be very flavourful. It is well to remember that an attractive plate to which some attention to colour has been given starts the digestive juices flowing in preparation for an enjoyable and well-digested meal. Many countries of the world use whole grains, prepared as suggested for beans, as the substantial part of their main meal. For those who feel they must have meat, it is suggested that relatively small portions be used and only with non-starch meals. In other words, omit bread, cake, pie, ice cream, macaroni, spaghetti, and the like, and use instead green vegetables, tomatoes, and other vegetables of a non-starchy type, with some fresh fruit for dessert. For cooking utensils use stainless steel, pottery or oven-glass, or even the old-fashioned iron pots.

Not only should food be wisely chosen, but it should also be wisely eaten. A sage old physician once said, 'Remember always that the stomach has no teeth.' Use the tools your body has, and chew your food thoroughly.

About Food

In this way, less food is needed, digestion improves, and the whole man benefits.

At this point it is well to note that organically grown food is far better for body building that food grown on soil depleted from over-cropping. Such soil is worn out and can give little value to anything grown on it. Properly prepared fertilizer must be used to build back into the soil the necessary minerals and trace minerals for the complete nourishment of the plants. The plants in turn can then provide man with the necessary elements in edible food form. Well-prepared soil not only makes proper plant growth possible, but it renders unnecessary the use of poisonous insecticides and sprays. Like people, plants provided with needed food elements naturally express health. In addition, the flavour of organically grown food is greatly improved, and the keeping qualities are vastly better.

Fertilizer for organic farming is usually obtained from 'compost', specially prepared from vegetable and mineral wastes, mixed with earth and water. These materials must be handled in such a way that the micro-organisms which break down the raw matter can do their work in the most efficient manner. This is a matter which warrants attention from the thoughtful reader interested in his own health and that of those around him. Therefore, we suggest that much useful information is available in *The Living Soil* by E. B. Balfour (*Faber*), *Soil and Sense* by Michael Graham (*Faber*), *Humus—and the Farmer*

120

About Food

by Friend Sykes (*Faber*), *Fertility Farming* by F. Newman Turner (*Faber*), *Soil Fertility, Renewal and Preservation* by Ehrenfried Pfeiffer (*Faber*), *Farming and Gardening in Health and Disease* by Sir Albert Howard (*Faber*), *Make Friends With Your Land* by Leonard Wickenden (*Devin Adair*) and *Pay Dirt* by J. I. Rodale (*Devin Adair*).

Since many books and charts given complete information regarding the minerals and vitamins in food, no attempt is made in this short chapter to give detailed information as to the values of the various articles of food. Besides, this is a treatise on health and not on disease. By using a well-rounded diet, the body will obtain in natural foods all that is necessary for a healthy and vigorous life.

Food is entertainment as well as building and replacement material. In order to obtain the most value from our meals we should eat them in an atmosphere of peace and harmony, never when we are over-tired, angry, or distressed. It is better to omit a meal than to eat when hurried or worried. A glass of fruit or vegetable juice is far better than a hurriedly bolted meal.

Even posture is important at meal time. In India people sit in the 'Happy Posture', which is cross-legged, on the ground. This is not practicable for Western people, but we should sit with our bodies well back in the chair, neither slumping, sitting on the edge of the chair, nor hunching over the table.

121

About Food

Make your meals simple. Do not use too many mixtures even of the good foods at any one meal. Prepare them attractively. Allow enough time for eating with a full measure of enjoyment. Follow these few simple rules and you will achieve that feeling of well-being which enables one to live rather than merely to exist.

Sample of Day's Meals

(May be amended to meet individual needs)

BREAKFAST *Summer* A large bowl of any one kind of fresh fruit. Cup of mint or camomile or any of the desired herb teas, with a little lemon and honey.

Winter On rising, juice of one lemon in large glass of hot water. Breakfast, not less than one-half hour later, may consist of a small dish of a good organically grown whole grain, a little cream and honey, with one or two slices of whole wheat bread, butter if desired, a few figs, dates, or apricots, and a cup of cereal coffee or herb tea.

LUNCH A large mixed salad of three or four vegetables with some cottage cheese or flaked nuts, fresh fruit for dessert. A variety from day to day will keep the meal interesting.

DINNER A large baked potato or savoury whole grains, a serving of either carrots, beets, or fresh green peas, also a serving of green beans or any green leafy vegetable cooked for as short a time as possible to make it tender. Baked or stewed fruit for dessert. Those who must have meat, can substitute this for the baked potato. Cereal coffee or delicious herb tea may end this simple but satisfying meal.

It is worth while experimenting for a few weeks with diets similar to that given above. Rarely does a great improvement in energy and vitality fail to appear.

List of Exercises